© Houghton Mifflin Harcourt Publishing Company • Cover Image Credits: (Bighorn Sheep)©Blaine Harrington III/ Alamy Images; (Watchman Peak, Utah) ©Russ Bishop/Alamy Images

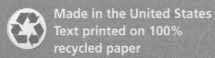

Made in the United States
Text printed on 100%
recycled paper

**Houghton
Mifflin
Harcourt**

Copyright © 2015 by Houghton Mifflin Harcourt Publishing Company

All rights reserved. No part of this work may be reproduced or transmitted in any form or by any means, electronic or mechanical, including photocopying or recording, or by any information storage and retrieval system, without the prior written permission of the copyright owner unless such copying is expressly permitted by federal copyright law. Requests for permission to make copies of any part of the work should be addressed to Houghton Mifflin Harcourt Publishing Company, Attn: Contracts, Copyrights, and Licensing, 9400 Southpark Center Loop, Orlando, Florida 32819-8647.

Common Core State Standards © Copyright 2010. National Governors Association Center for Best Practices and Council of Chief State School Officers. All rights reserved.

This product is not sponsored or endorsed by the Common Core State Standards Initiative of the National Governors Association Center for Best Practices and the Council of Chief State School Officers.

Printed in the U.S.A.

ISBN 978-0-544-34253-8

14 15 16 17   0928   22 21 20 19 18

4500736971   C D E F G

If you have received these materials as examination copies free of charge, Houghton Mifflin Harcourt Publishing Company retains title to the materials and they may not be resold. Resale of examination copies is strictly prohibited.

Possession of this publication in print format does not entitle users to convert this publication, or any portion of it, into electronic format.

Dear Students and Families,

Welcome to **Go Math!**, Grade 6! In this exciting mathematics program, there are hands-on activities to do and real-world problems to solve. Best of all, you will write your ideas and answers right in your book. In **Go Math!**, writing and drawing on the pages helps you think deeply about what you are learning, and you will really understand math!

By the way, all of the pages in your **Go Math!** book are made using recycled paper. We wanted you to know that you can Go Green with **Go Math!**

Sincerely,

The Authors

Made in the United States
Text printed on 100% recycled paper

© Houghton Mifflin Harcourt Publishing Company • Image Credits: (bg) ©Sankar Salvady/Flickr/Getty Images; (t) ©Blaine Harrington III/Alamy Images; (c) ©Don Johnston/All Canada Photos/Getty Images; (b) ©Erich Kuchling/Westend61/Corbis

GO MATH!

Authors

Juli K. Dixon, Ph.D.
Professor, Mathematics Education
University of Central Florida
Orlando, Florida

Edward B. Burger, Ph.D.
President, Southwestern University
Georgetown, Texas

Steven J. Leinwand
Principal Research Analyst
American Institutes for
    Research (AIR)
Washington, D.C.

Contributor

Rena Petrello
Professor, Mathematics
Moorpark College
Moorpark, California

Matthew R. Larson, Ph.D.
K-12 Curriculum Specialist for
    Mathematics
Lincoln Public Schools
Lincoln, Nebraska

Martha E. Sandoval-Martinez
Math Instructor
El Camino College
Torrance, California

English Language Learners Consultant

Elizabeth Jiménez
CEO, GEMAS Consulting
Professional Expert on English
    Learner Education
Bilingual Education and
    Dual Language
Pomona, California

© Houghton Mifflin Harcourt Publishing Company • Image Credits: (bg) ©Russ Bishop/Alamy Images P (t) ©Richard Wear/Design Pics/Corbis

Geometry and Statistics

Critical Area Solve real-world and mathematical problems involving area, surface area, and volume; and developing understanding of statistical thinking

13 Variability and Data Distributions · **705**

COMMON CORE STATE STANDARDS

6.SP Statistics and Probability
Cluster A Develop understanding of statistical variability.
6.SP.A.2, 6.SP.A.3
Cluster B Summarize and describe distributions.
6.SP.B.4, 6.SP.B.5c, 6.SP.B.5d

© Houghton Mifflin Harcourt Publishing Company

Critical Area

GO DIGITAL

Go online! Your math lessons are interactive. Use *i*Tools, Animated Math Models, the Multimedia *e*Glossary, and more.

Chapter 13 Overview

In this chapter, you will explore and discover answers to the following **Essential Questions**:

- How can you describe the shape of a data set using graphs, measures of center, and measures of variability?
- How do you calculate the different measures of center?
- How do you calculate the different measures of variability?

Personal Math Trainer
Online Assessment and Intervention

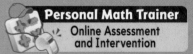

CRITICAL AREA REVIEW PROJECT THE RIGHT PRICE: *www.thinkcentral.com*

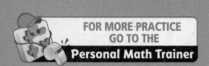

FOR MORE PRACTICE
GO TO THE
Personal Math Trainer

Practice and Homework

Lesson Check and
Spiral Review in
every lesson

© Houghton Mifflin

Variability and Data Distributions

Show What You Know

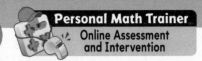

Personal Math Trainer
Online Assessment and Intervention

Check your understanding of important skills.

Name _____

▶ **Place the First Digit** Tell where to place the first digit. Then divide. (4.NBT.B.6)

1. $4\overline{)872}$ _____ place

2. $8\overline{)256}$ _____ place

▶ **Order of Operations** Evaluate the expression. (6.EE.A.2)

3. $9 + 4 \times 8$

_____

4. $2 \times 7 + 5$

_____

5. $6 \div (3 - 2)$

_____

6. $(12 - 3^2) \times 5$

_____

7. $2^3 \times (22 \div 2)$

_____

8. $(8 - 2)^2 - 9$

_____

9. $(9 - 2^3) + 8$

_____

10. $(27 + 9) \div 3$

_____

▶ **Mean** Find the mean for the set of data. (6.SP.B.5c)

11. 285, 420, 345, 390 _____

12. 0.2, 0.23, 0.16, 0.21, 0.2 _____

13. $33, $48, $55, $52 _____

14. 8.1, 7.2, 8.4 _____

Raina watched two of her friends play a game of darts.
She has to pick one of them to be her partner in a
tournament. Help Raina figure out which of her friends
is a more consistent dart player.

_____

_____

Dart Scores						
Hector	15	5	7	19	3	19
Marin	12	10	11	11	10	14

© Houghton Mifflin Harcourt Publishing Company • Image Credits: (br) ©PhotoDisc/Getty Images

Vocabulary Builder

▶ Visualize It

Sort the review words into the chart.

Measures of Center

How Do I Find It?

Find the sum of all the data values and divide the sum by the number of data values.

Order the data and find the middle value or the mean of the two middle values if the number of values is even.

Find the data value(s) that occurs most often.

Review Words

histogram

mean

median

mode

Preview Words

box plot

lower quartile

interquartile range

measure of variability

range

upper quartile

▶ Understand Vocabulary

Complete the sentences using the preview words.

1. The median of the upper half of a data set is the

   _____.

2. The _____ is the difference
   between the greatest value and the least value in a data set.

3. A(n) _____ is a graph that shows the median,
   quartiles, least value, and greatest value of a data set.

4. A data set's _____ is the difference between
   its upper and lower quartiles.

5. You can describe how spread out a set of data is using a(n)

   _____.

• **Interactive Student Edition**
• **Multimedia eGlossary**

© Houghton Mifflin Harcourt Publishing Company

Chapter 13 Vocabulary

box plot

diagrama de caja

8

interquartile range

rango intercuartil

45

lower quartile

primer cuartil

54

mean absolute deviation

desviación absoluta respecto a la media

56

measure of variability

medida de dispersión

58

median

mediana

59

range

rango

85

upper quartile

tercer cuartil

107

The difference between the upper and lower quartiles of a data set

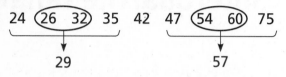

24 (26 32) 35   42   47 (54 60) 75

       29   57

  lower quartile   upper quartile

The interquartile range is 57 − 29 = 28.

© Houghton Mifflin Harcourt Publishing Company

A graph that shows how data are distributed using the median, quartiles, least value, and greatest value

Example:

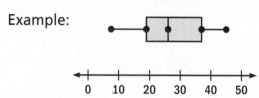

© Houghton Mifflin Harcourt Publishing Company

The mean of the distances from each data value in a set to the mean of the set

© Houghton Mifflin Harcourt Publishing Company

The median of the lower half of a data set

Example:

24 (26 32) 35   42   47 (54 60) 75

       29   57

  lower quartile   upper quartile

© Houghton Mifflin Harcourt Publishing Company

The middle value when a data set is written in order from least to greatest, or the mean of the two middle values when there is an even number of items

Example:

8, 17, 21, 23, (26) 29, 34, 40, 45

© Houghton Mifflin Harcourt Publishing Company

A single value used to describe how the values in a data set are spread out

Examples: range, interquartile range, mean absolute deviation

© Houghton Mifflin Harcourt Publishing Company

The median of the upper half of a data set

Example:

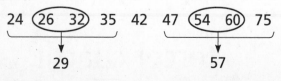

24 (26 32) 35   42   47 (54 60) 75

       29   57

  lower quartile   upper quartile

© Houghton Mifflin Harcourt Publishing Company

The difference between the greatest and least numbers in a data set

Example: The range of the data set 60, 35, 22, 46, 81, 39 is 81 − 22 = 59.

© Houghton Mifflin Harcourt Publishing Company

Matchup

For 2–3 players

Materials
- 1 set of word cards

How to Play

1. Place the cards face-down on a table in even rows. Take turns to play.

2. Choose two cards and turn them face-up.
   - If the cards show a word and its meaning, it's a match. Keep the pair and take another turn.
   - If the cards do not match, turn them over again.

3. The game is over when all cards have been matched. The players count their pairs. The player with the most pairs wins.

Word Box

box plot

interquartile range

lower quartile

mean absolute deviation

measure of variability

median

range

upper quartile

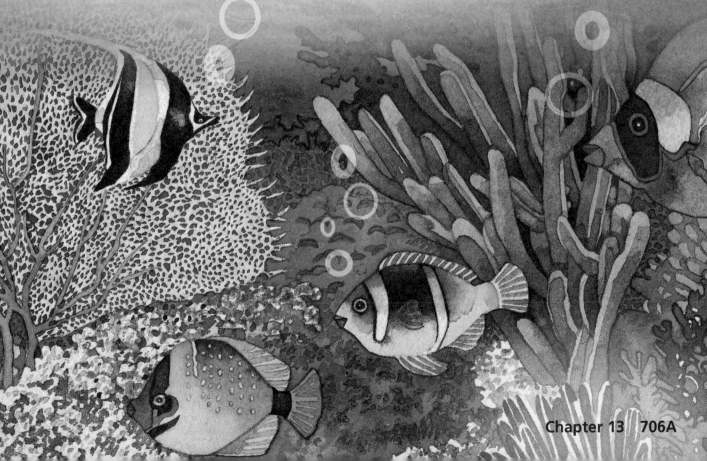

© Houghton Mifflin Harcourt Publishing Company • Image Credits: (b) ©PhotoDisc/Getty Images; (t) ©PhotoDisc/Getty Images; (b) ©Bobbi Tull/PhotoDisc/Getty Images

The Write Way

Reflect

Choose one idea. Write about it.

- Summarize the information represented in a box plot.
- Explain in your own words what the *mean absolute deviation* is.
- Joe says that the median for a set of data is a measure of variability. Yossi says that the range is. Identify who is correct and explain why.
- Describe how to calculate the interquartile range for a set of data.

© Houghton Mifflin Harcourt Publishing Company • Image Credits: (bg) ©Comstock/Getty Images; (t) ©PhotoDisc/Getty Images; (b) ©Bobbi Tull/PhotoDisc/Getty Images

Name _____

Patterns in Data

Essential Question How can you describe overall patterns in a data set?

Common Core **Statistics and Probability—6.SP.B.5c**
Also 6.SP.A.2
MATHEMATICAL PRACTICES
MP7, MP8

CONNECT Seeing data sets in graphs, such as dot plots and histograms, can help you find and understand patterns in the data.

Unlock the Problem

Many lakes and ponds contain freshwater fish species such as bass, pike, bluegill, and trout. Jacob and his friends went fishing at a nearby lake. The dot plot shows the sizes of the fish that the friends caught. The lengths are rounded to the nearest inch. What patterns do you see in the data?

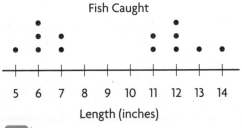

Fish Caught

Length (inches)
5 6 7 8 9 10 11 12 13 14

- Circle any spaces with no data.
- Place a box around any groups of data.

Analyze the dot plot.

A *gap* is an interval that contains no data.

Does the dot plot contain any gaps? If so, where? _____

A *cluster* is a group of data points that lie within a small interval.

There is a cluster from _____ to _____ and another cluster from _____ to _____.

So, there were no fish from _____ to _____ inches long,

and there were two clusters of fish measuring from _____

to _____ inches long and from _____ to _____ inches long.

Math Talk

MATHEMATICAL PRACTICES ⑦

Look for Structure What is the mode(s) of the data? Explain how you know.

1. Summarize the information shown in the dot plot.

_____

_____

2. **MATHEMATICAL PRACTICE ⑧ Draw Conclusions** What conclusion can you draw about why the data might have this pattern?

_____

_____

© Houghton Mifflin Harcourt Publishing Company • Image Credits: (r) ©WILDLIFE GmbH/Alamy Images

You can also analyze patterns in data that are displayed in histograms. Some data sets have symmetry about a peak, while others do not.

Example   Analyze a histogram.

Erica made this histogram to show the weights of the pumpkins grown at her father's farm in October. What patterns do you see in the data?

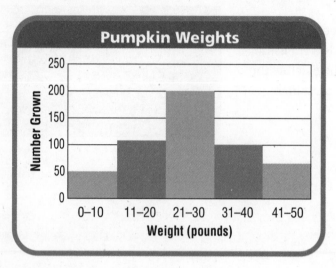

STEP 1 Identify any peaks in the data.

The histogram has _____ peak(s).

The interval representing the greatest number of pumpkins is for

weights between _____ and _____ pounds.

_____

STEP 2 Describe how the data changes across the intervals.

The number of pumpkins increases from 0 to _____ pounds

and _____ from 31 to 50 pounds.

_____

STEP 3 Describe any symmetry the graph has.

If I draw a vertical line through the interval for _____ to

_____ pounds, the left and right sides of the histogram are very

close to being mirror images. The histogram _____ symmetry.

> **Math Idea**
>
> A graph can have symmetry if you can draw a line through it so that the two sides of the graph are almost mirror images of each other.

So, the data values increase to one peak in the interval for _____ to

_____ pounds and then decrease. The data set _____ symmetry about the peak.

© Houghton Mifflin Harcourt Publishing Company • Image Credits: (r) ©Siaukia/Alamy Images

Name _____

For 1–3, use the dot plot.

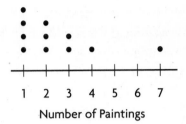

1. The dot plot shows the number of paintings students in the art club displayed at the art show. Does the dot plot contain any gaps?

   If so, where? _____

   Number of Paintings

2. Identify any clusters in the data.

   _____

3. Summarize the information in the dot plot.

   _____

   _____

On Your Own

4. **GO DEEPER** What patterns do you see in the histogram data?

   _____

   _____

   _____

   _____

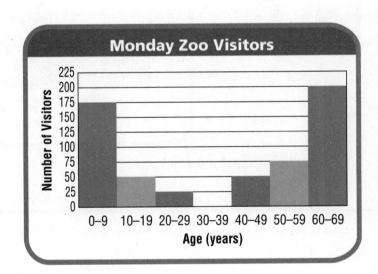

5. **THINK SMARTER** The dot plot shows the number of errors made by a baseball team in the first 16 games of the season. For numbers 5a-5e, choose Yes or No to indicate whether the statement is correct.

   Errors per Game

   5a. There is a gap from 4 to 5.   ○ Yes   ○ No

   5b. There is a peak at 0.   ○ Yes   ○ No

   5c. The dot plot has symmetry.   ○ Yes   ○ No

   5d. There are two modes.   ○ Yes   ○ No

   5e. There is one cluster.   ○ Yes   ○ No

© Houghton Mifflin Harcourt Publishing Company

Connect to Science

Big Cats

There are 41 species of cats living in the world today. Wild cats live in places as different as deserts and the cold forests of Siberia, and they come in many sizes. Siberian tigers may be as long as 9 feet and weigh over 2,000 pounds, while bobcats are often just 2 to 3 feet long and weigh between 15 and 30 pounds.

You can find bobcats in many zoos in the United States. The histogram below shows the weights of several bobcats. The weights are rounded to the nearest pound.

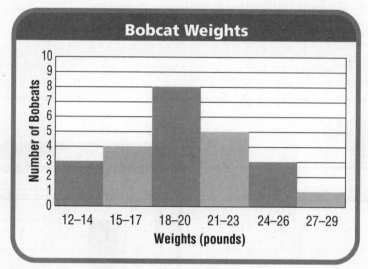

Use the histogram for 6 and 7.

6. **MATHEMATICAL PRACTICE 7 Look for a Pattern** Describe the overall shape of the histogram.

_____

_____

_____

_____

7. **THINK SMARTER** **Sense or Nonsense?** Sunny says that the graph might have a different shape if it was redrawn as a bar graph with one bar for each number of pounds. Is Sunny's statement sense or nonsense? Explain.

_____

_____

_____

_____

© Houghton Mifflin Harcourt Publishing Company • Image Credits: (t) ©Arco Images GmbH/Alamy Images

Patterns in Data

COMMON CORE STANDARD—6.SP.B.5c
Summarize and describe distributions.

For 1–2, use the dot plot.

1. The dot plot shows the number of omelets ordered at Paul's Restaurant each day. Does the dot plot contain any gaps?

   _____ **Yes; from 12 to 13, and at 17** _____

2. Identify any clusters in the data.

   _____

   _____

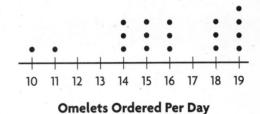

Omelets Ordered Per Day

For 3–4, use the histogram.

3. The histogram shows the number of people that visited a local shop each day in January. How many peaks does the histogram have?

   _____

4. Describe how the data values change across the intervals.

   _____

   _____

   _____

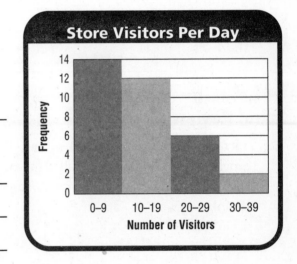

Problem Solving *Real World*

5. Look at the dot plot at the right. Does the graph have symmetry? Explain.

   _____

   _____

   _____

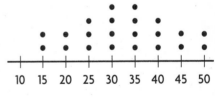

Gift Cards Purchased This Week

6. **WRITE** ▶*Math* A histogram that shows the ages of students at a library has intervals 1–5, 6–10, 11–15, 16–20, and 21–25. There is a peak at 11–15 years and the graph is symmetric. Sketch what the histogram could look like and describe the patterns you see in the data.

   _____

© Houghton Mifflin Harcourt Publishing Company

Lesson Check (6.SP.B.5c)

1. What interval in the histogram has the greatest frequency?

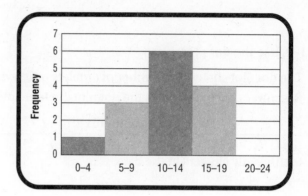

_____

2. Meg makes a dot plot for the data 9, 9, 4, 5, 5, 3, 4, 5, 3, 8, 8, 5. Where does a gap occur?

_____

Spiral Review (6.G.A.2, 6.SP.B.4, 6.SP.B.5c)

3. A rectangular fish tank is 20 inches long, 12 inches wide, and 20 inches tall. If the tank is filled halfway with water, how much water is in the tank?

_____

4. Look at the histogram below. How many students scored an 81 or higher on the math test?

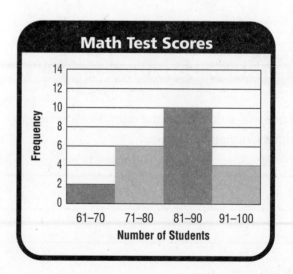

5. The Little League coach uses a radar gun to measure the speed of several of Kyle's baseball pitches. The speeds, in miles per hour, are 52, 48, 63, 47, 47. What is the median of Kyle's pitch speeds?

_____   _____

© Houghton Mifflin Harcourt Publishing Company

**FOR MORE PRACTICE
GO TO THE
Personal Math Trainer**

Name _____

Box Plots

Essential Question How can you use box plots to display data?

 Common Core Statistics and Probability—
6.SP.B.4
MATHEMATICAL PRACTICES
MP1, MP3, MP6

The median is the middle value, or the mean of the two middle values, when data is written in order. The **lower quartile** is the median of the lower half of a data set, and the **upper quartile** is the median of the upper half of a data set.

⚷ Unlock the Problem (Real World)

In 1885, a pair of jeans cost $1.50. Today, the cost of jeans varies greatly. The chart lists the prices of jeans at several different stores. What are the median, lower quartile, and upper quartile of the data?

Prices of Jeans								
$35	$28	$42	$50	$24	$75	$47	$32	$60

 Find the median, lower quartile, and upper quartile.

STEP 1 Order the numbers from least to greatest.

$24 $28 $32 $35 $42 $47 $50 $60 $75

STEP 2 Circle the middle number, or the median.

The median is $ _____.

STEP 3 Calculate the upper and lower quartiles.

Find the median of each half of the data set.

Think: If a data set has an even number of values, the median is the mean of the two middle values.

> **ERROR Alert**
>
> When a data set has an odd number of values, do not include the median when finding the lower and upper quartiles.

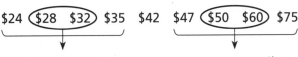

$24 ⟨$28 $32⟩ $35 $42 $47 ⟨$50 $60⟩ $75

    lower quartile   upper quartile

$$\frac{\$28 + \$32}{2} = \frac{\$\_\_\_\_}{2} = \$\_\_\_\_ \qquad \frac{\$\_\_\_\_ + \$\_\_\_\_}{2} = \frac{\$\_\_\_\_}{2} = \$\_\_\_\_$$

So, the median is $ _____, the lower quartile is $ _____, and the

upper quartile is $ _____.

© Houghton Mifflin Harcourt Publishing Company • Image Credits: (r) ©Corbis

A **box plot** is a type of graph that shows how data are distributed by using the least value, the lower quartile, the median, the upper quartile, and the greatest value. Below is a box plot showing the data for jean prices from the previous page.

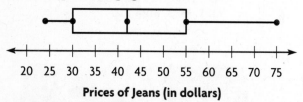

Prices of Jeans (in dollars)

 Example   Make a box plot.

The data set below represents the ages of the top ten finishers in a 5K race. Use the data to make a box plot.

Ages of Top 10 Runners (in years)									
33	18	21	23	35	19	38	30	23	25

STEP 1 Order the data from least to greatest. Then find the median and the lower and upper quartiles.

18, _____, _____, _____, _____, _____, _____, _____, _____, _____

Median = $\dfrac{\boxed{} + \boxed{}}{2}$ = _____ years

Lower quartile = _____ years   The lower quartile is the median of the lower half of the data set, which includes the values from 18 to 23.

Upper quartile = _____ years   The upper quartile is the median of the upper half of the data set, which includes the values from 25 to 38.

STEP 2 Draw a number line. Above the number line, plot a point for the least value, the lower quartile, the median, the upper quartile, and the greatest value.

Ages of Top Ten Runners

STEP 3 Draw a box from the lower to upper quartile. Inside the box, draw a vertical line segment through the median. Then draw line segments from the box to the least and greatest values.

Math Talk

MATHEMATICAL PRACTICES ⑥

Describe the steps for making a box plot.

• **MATHEMATICAL PRACTICE ⑥ Explain** Would the box plot change if the data point for 38 years were replaced with 40 years? Explain.

© Houghton Mifflin Harcourt Publishing Company • Image Credits: (t) ©Image Source/Getty Images

Name _____

Find the median, lower quartile, and upper quartile of the data.

1. the scores of 11 students on a geography quiz:
87, 72, 80, 95, 86, 80, 78, 92, 88, 76, 90

   Order the data from least to greatest.   72, 76, 78, 80, 80, 86, 87, 88, 90, 92, 95

   median: _____   lower quartile: _____   upper quartile: _____

2. the lengths, in seconds, of 9 videos posted online:
50, 46, 51, 60, 62, 50, 65, 48, 53

   median: _____   lower quartile: _____   upper quartile: _____

3. Make a box plot to display the data set in Exercise 2.

Lengths of Online Videos (seconds)

On Your Own

Math Talk   MATHEMATICAL PRACTICES ⑥

Compare How are box plots and dot plots similar? How are they different?

Find the median, lower quartile, and upper quartile of the data.

4. 13, 24, 37, 25, 56, 49, 43, 20, 24

   median: _____

   lower quartile: _____

   upper quartile: _____

5. 61, 23, 49, 60, 83, 56, 51, 64, 84, 27

   median: _____

   lower quartile: _____

   upper quartile: _____

6. The chart shows the height of trees in a park.
Display the data in a box plot.

Tree Heights (feet)											
8	12	20	30	25	18	18	8	10	28	26	29

Tree Heights (feet)

7. **MATHEMATICAL PRACTICE ①** **Analyze** Eric made this box plot for the data set below. Explain his error.

Number of Books Read								
5	13	22	8	31	37	25	24	10

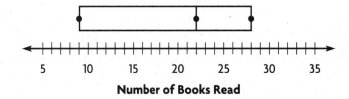

Number of Books Read

© Houghton Mifflin Harcourt Publishing Company

MATHEMATICAL PRACTICES   MODEL · REASON · MAKE SENSE

Problem Solving · Applications

THINK SMARTER   **Pose a Problem**

8. The box plots show the number of flights delayed per day
for two different airlines. Which data set is more spread out?

Find the distance between the least and greatest values for
each data set.

Airline A: greatest value − least value =

_____ − _____ = _____

Airline B: greatest value − least value =

_____ − _____ = _____

So, the data for _____ is more spread out.

Write a new problem that can be solved using the
data in the box plots.

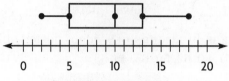

Flights Delayed: Airline A

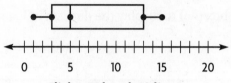

Flights Delayed: Airline B

Pose a Problem

_____

_____

_____

_____

_____

Solve Your Problem

9. *THINK SMARTER*   The data set shows the cost of the dinner specials at a
restaurant on Friday night.

Cost of Dinner Specials ($)										
30	24	24	16	24	25	19	28	18	19	26

The median is

19.
24.
25.

The lower quartile is

16.
18.
19.

The upper quartile is

26.
28.
30.

© Houghton Mifflin Harcourt Publishing Company

Box Plots

Common Core **COMMON CORE STANDARD—6.SP.B.4**
Summarize and describe distributions.

Find the median, lower quartile, and upper quartile of the data.

1. the amounts of juice in 12 glasses, in fluid ounces:

   11, 8, 4, 9, 12, 14, 9, 16, 15, 11, 10, 7

   Order the data from least to greatest: 4, 7, 8, 9, 9, 10, 11, 11, 12, 14, 15, 16

   median: ___10.5___   lower quartile: ___8.5___   upper quartile: ___13___

2. the lengths of 10 pencils, in centimeters:

   18, 15, 4, 9, 14, 17, 16, 6, 8, 10

   median: _____   lower quartile: _____   upper quartile: _____

3. Make a box plot to display the data set in Exercise 2.

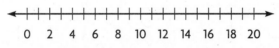

Lengths of Pencils (centimeters)

4. The numbers of students on several teams are 9, 4, 5, 10, 11, 9, 8, and 6. Make a box plot for the data.

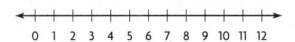

Number of Students on a Team

Problem Solving (Real World)

5. The amounts spent at a gift shop today are $19, $30, $28, $22, $20, $26, and $26. What is the median? What is the lower quartile?

   _____

6. The weights of six puppies in ounces are 8, 5, 7, 5, 6, and 9. What is the upper quartile of the data?

   _____

7. **WRITE** ▸*Math* Draw a box plot to display this data: 81, 22, 34, 55, 76, 20, 56.

   _____

   _____

© Houghton Mifflin Harcourt Publishing Company

Lesson Check (6.SP.B.4)

1. The values in a data set are 15, 7, 11, 12, 6, 3, 10, and 6. Where would you draw the box in a box plot for the data?

2. What is the lower quartile of the following data set?

22, 27, 14, 21, 22, 26, 18

Spiral Review (6.SP.A.1, 6.SP.B.5c, 6.SP.B.5d)

3. Jenn says that "What is the average number of school lunches bought per day?" is a statistical question. Lisa says that "How many lunches did Mark buy this week?" is a statistical question. Who is NOT correct?

4. The prices of several chairs are $89, $76, $81, $91, $88, and $70. What is the mean of the chair prices?

5. By how much does the mean of the following data set change if the outlier is removed?

13, 19, 16, 40, 12

6. Where in the dot plot does a cluster occur?

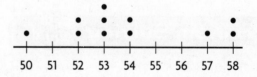

© Houghton Mifflin Harcourt Publishing Company

FOR MORE PRACTICE GO TO THE Personal Math Trainer

Name _____

Mean Absolute Deviation

Essential Question How do you calculate the mean absolute deviation of a data set?

Common Core — Statistics and Probability—6.SP.B.5c

MATHEMATICAL PRACTICES
MP2, MP3, MP4, MP6

One way to describe a set of data is with the mean. However, two data sets may have the same mean but look very different when graphed. When interpreting data sets, it is important to consider how far away the data values are from the mean.

Investigate

Materials ▪ counters, large number line from 0–10

The number of magazine subscriptions sold by two teams of students for a drama club fundraiser is shown below. The mean number of subscriptions for each team is 4.

Team A				
3	3	4	5	5

Team B				
0	1	4	7	8

A. Make a dot plot of each data set using counters for the dots. Draw a vertical line through the mean.

B. Count to find the distance between each counter and the mean. Write the distance underneath each counter.

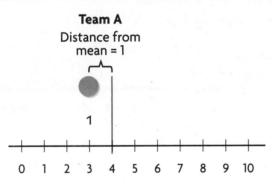

Team A
Distance from
mean = 1

1

0 1 2 3 4 5 6 7 8 9 10

C. Find the mean of the distances for each data set.

Team A

$$\frac{1 + \quad + \quad + \quad + \quad}{5} = \frac{\quad}{5} = \quad$$

Team B

$$\frac{\quad + \quad + \quad + \quad + \quad}{\quad} = \frac{\quad}{\quad} = \quad$$

© Houghton Mifflin Harcourt Publishing Company • Image Credits: (r) ©Houghton Mifflin Harcourt

Draw Conclusions

1. THINK SMARTER Which data set, Team A or B, looks more spread
    out in your dot plots? Which data set had a greater average distance
    from the mean? Explain how these two facts are connected.

    _____

    _____

2. MATHEMATICAL PRACTICE ② **Reason Quantitatively** The table shows
    the average distance from the mean for the heights of
    players on two basketball teams. Tell which set of heights
    is more spread out. Explain how you know.

    _____

    _____

Heights of Players	
Team	Average Distance from Mean (in.)
Chargers	2.8
Wolverines	1.5

Make Connections

The mean of the distances of data values from the mean of the data set
is called the **mean absolute deviation**. As you learned in the Investigation,
mean absolute deviation is a way of describing how spread out a data set is.

The dot plot shows the ages of gymnasts registered for the school team.
The mean of the ages is 10. Find the mean absolute deviation of the data.

STEP 1 Label each dot with its distance from the mean.

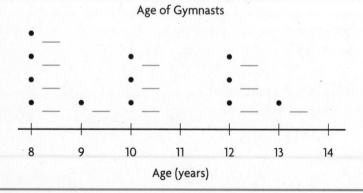

Age of Gymnasts

Math Talk

MATHEMATICAL PRACTICES ③

Apply Is it possible for the mean
absolute deviation of a data set to
be zero? Explain.

STEP 2 Find the mean of the distances.

$$\frac{\boxed{}+\boxed{}+\boxed{}+\boxed{}+\boxed{}+\boxed{}+\boxed{}+\boxed{}+\boxed{}+\boxed{}+\boxed{}}{\boxed{}} = \underline{\quad} = \boxed{}$$

So, the mean absolute deviation of the data is _____ years.

© Houghton Mifflin Harcourt Publishing Company

Name _____

Use counters, a dot plot, or *i*Tools to find the mean absolute deviation of the data.

1. Find the mean absolute deviation for both data sets. Explain which data set is more spread out.

the number of laps Shawna swam on 5 different days:

5, 6, 6, 8, 10

mean = 7

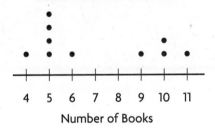

$$\frac{2 + \quad + \quad + \quad + \quad}{} = \frac{}{} = $$

mean absolute deviation = _____ laps

the number of laps Lara swam on 5 different days:

1, 3, 7, 11, 13

mean = 7

mean absolute deviation = _____ laps

The data set of _____ laps is more spread out because the mean

absolute deviation of her data is _____.

Use the dot plot to find the mean absolute deviation of the data.

2. mean = 7 books

Books Read Each Semester

```
            •
            •
            •          •
   •    •   •      •    •    •
 --+----+----+----+----+----+----+----+--
   4    5    6    7    8    9   10   11
           Number of Books
```

mean absolute deviation = _____

3. mean = 29 pounds

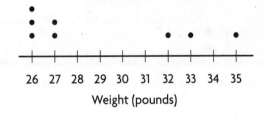

Packages Shipped on Tuesday

```
   •
   •    •
   •    •              •    •         •
 --+----+----+----+----+----+----+----+----+----+--
  26   27   28   29   30   31   32   33   34   35
              Weight (pounds)
```

mean absolute deviation = _____

4. **WRITE** ▸*Math* The mean absolute deviation of the number of daily visits to Scott's website for February is 167.7. In March, the absolute mean deviation is 235.9. In which month did the number of visits to Scott's website vary more? Explain how you know.

_____

5. **MATHEMATICAL PRACTICE ④** **Write an Inequality** **Algebra** In April, the data for Scott's website visits are less spread out than they were in February. Use *a* to represent the mean absolute deviation for April. Write an inequality to describe the possible values of *a*.

_____

© Houghton Mifflin Harcourt Publishing Company

Problem Solving • Applications

6. **GO DEEPER** Use the table.

Days of Precipitation											
Jan	Feb	Mar	Apr	May	Jun	Jul	Aug	Sep	Oct	Nov	Dec
10	12	13	18	10	8	7	6	16	14	8	10

The mean of the data is 11. What is the mean absolute deviation of the data?

_____

7. **THINK SMARTER** Suppose all of the players on a basketball team had the same height. Explain how you could use reasoning to find the mean absolute deviation of the players' heights.

_____

_____

8. **MATHEMATICAL PRACTICE ⑥ Explain** Tell how an outlier that is much greater than the mean would affect the mean absolute deviation of the data set. Explain your reasoning.

_____

_____

_____

9. **THINK SMARTER** The data set shows the number of soccer goals scored by players in 3 games.

For numbers 9a–9c, choose Yes or No to indicate whether the statement is correct.

Number of Goals Scored			
Player A	1	2	1
Player B	2	2	2
Player C	3	2	1

9a.   The mean absolute deviation of Player A is 1.   ○ Yes   ○ No

9b.   The mean absolute deviation of Player B is 0.   ○ Yes   ○ No

9c.   The mean absolute deviation of Player C is greater than the mean absolute deviation of Player A.   ○ Yes   ○ No

© Houghton Mifflin Harcourt Publishing Company • Image Credits: (t) ©Johner Images/Alamy Images

Mean Absolute Deviation

Use counters and a dot plot to find the mean absolute deviation of the data.

Common Core

COMMON CORE STANDARD—6.SP.B.5c
Summarize and describe distributions.

1. the number of hours Maggie spent practicing soccer for 4 different weeks:

   9, 6, 6, 7

   mean = 7 hours

   $\dfrac{2 + 1 + 1 + 0}{4} = \dfrac{4}{4} = 1$

   mean absolute deviation = _____**1 hour**_____

2. the heights of 7 people in inches:

   60, 64, 58, 60, 70, 71, 65

   mean = 64 inches

   mean absolute deviation = _____

Use the dot plot to find the mean absolute deviation of the data.

3. mean = 10

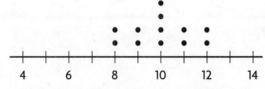

   Ages of Students in Dance Class

   mean absolute deviation = _____

4. mean = 8

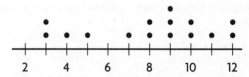

   Weekly Hours Spent Doing Homework

   mean absolute deviation = _____

Problem Solving · Real World

5. In science class, Troy found the mass, in grams, of 6 samples to be 10, 12, 7, 8, 5, and 6. What is the mean absolute deviation?

6. Five recorded temperatures are 71°F, 64°F, 72°F, 81°F, and 67°F. What is the mean absolute deviation?

   _____

   _____

7. **WRITE** ▸*Math* Make a dot plot of the following data: 10, 10, 11, 12, 12, 13, 13, 15. Use the dot plot to find the mean absolute deviation.

   _____

   _____

© Houghton Mifflin Harcourt Publishing Company

Lesson Check (6.SP.B.5c)

1. Six test grades are 86, 88, 92, 90, 82, and 84. The mean of the data is 87. What is the mean absolute deviation?

2. Eight heights in inches are 42, 36, 44, 46, 48, 42, 48, and 46. The mean of the data is 44. What is the mean absolute deviation?

Spiral Review (6.G.A.2, 6.SP.B.4)

3. What is the volume of a rectangular prism with dimensions 4 meters, $1\frac{1}{2}$ meters, and 5 meters?

4. Carrie is making a frequency table showing the number of miles she walked each day during the 30 days of September. What value should she write in the Frequency column for 9 to 11 miles?

Carrie's Daily Walks	
Number of Miles	Frequency
0–2	17
3–5	8
6–8	4
9–11	?

5. The following data shows the number of laps each student completed. What number of laps is the mode?

9, 6, 7, 8, 5, 1, 8, 10

6. What is the upper quartile of the following data?

43, 48, 55, 50, 58, 49, 38, 42, 50

© Houghton Mifflin Harcourt Publishing Company

FOR MORE PRACTICE
GO TO THE
Personal Math Trainer

Name _____

Measures of Variability

Essential Question How can you summarize a data set by using range, interquartile range, and mean absolute deviation?

Common Core Statistics and Probability— 6.SP.B.5c
Also 6.SP.A.2, 6.SP.A.3
MATHEMATICAL PRACTICES
MP6, MP7

CONNECT A **measure of variability** is a single value used to describe how spread out a set of data values are. The mean absolute deviation is a measure of variability.

 Unlock the Problem Real World

In gym class, the students recorded how far they could jump. The data set below gives the distances in inches that Manuel jumped. What is the mean absolute deviation of the data set?

Manuel's Jumps (in inches)					
54	58	56	59	60	55

🔑 **Find the mean absolute deviation.**

STEP 1 Find the mean of the data set.

Add the data values and divide the sum by the number of data values.

$54 + \underline{} + \underline{} + \underline{} + \underline{} + \underline{} = \dfrac{}{} = $

The mean of the data set is _____ inches.

STEP 2 Find the distance of each data value from the mean.

Subtract the lesser value from the greater value.

Data Value	Subtract (Mean = 57)	Distance between data value and the mean
54	57 − 54 =	3
58	58 − 57 =	
56	57 − 56 =	
59	59 − 57 =	
60	60 − 57 =	
55	57 − 55 =	

Total of distances from the mean:

STEP 3 Add the distances.

STEP 4 Find the mean of the distances.

Divide the sum of the distances by the number of data values.

_____ ÷ 6 = _____

So, the mean absolute deviation of the data is _____ inches.

Math Talk **MATHEMATICAL PRACTICES ⑦**

Look for Structure Give an example of a data set that has a small mean absolute deviation. Explain how you know that the mean absolute deviation is small without doing any calculations.

© Houghton Mifflin Harcourt Publishing Company • Image Credits: (tr) ©Jim Lane/Alamy Images

Range is the difference between the greatest value and the least value in a data set. **Interquartile range** is the difference between the upper quartile and the lower quartile of a data set. Range and interquartile range are also measures of variability.

Example Use the range and interquartile range to compare the data sets.

The box plots show the price in dollars of the handheld game players at two different electronic stores. Find the range and interquartile range for each data set. Then compare the variability of the prices of the handheld game players at the two stores.

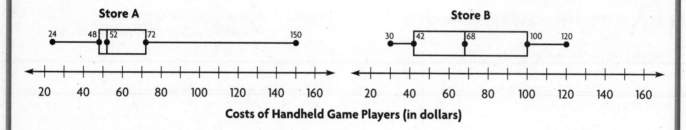

Store A

24   48 52   72   150

Store B

30   42   68   100   120

20   40   60   80   100   120   140   160

20   40   60   80   100   120   140   160

Costs of Handheld Game Players (in dollars)

	STORE A	**STORE B**
Calculate the range. Find the difference between the greatest and least values.	150 − 24 = _____ The range for Store A is _____.	120 − _____ = _____ The range for Store B is _____.
Calculate the interquartile range. Find the difference between the upper quartile and lower quartile.	72 − 48 = _____ The interquartile range for Store A is _____.	100 − _____ = _____ The interquartile range for Store B is _____.

So, Store A has a greater _____, but

Store B has a greater _____.

Math Talk

MATHEMATICAL PRACTICES ⑥

Compare Explain how range and interquartile range are alike and how they are different.

© Houghton Mifflin Harcourt Publishing Company • Image Credits: (tr) ©Randy Faris/Corbis

Name _____

 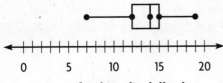
✓ **1.** Find the range and interquartile range of the data in the box plot.

0   5   10   15   20

Cost of T-shirts (in dollars)

For the range, find the difference between the greatest and least values.

_____ – _____ = _____

range: $ _____

For the interquartile range, find the difference between the upper and lower quartiles.

_____ – _____ = _____

interquartile range: $ _____

Practice: Copy and Solve Find the mean absolute deviation for the data set.

✓ **2.** heights in inches of several tomato plants:

16, 18, 18, 20, 17, 20, 18, 17

mean absolute deviation: _____

3. times in seconds for students to run one lap:

68, 60, 52, 40, 64, 40

mean absolute deviation: _____

On Your Own

Math Talk — MATHEMATICAL PRACTICES ⑥

Explain how you can find the mean absolute deviation of a data set.

Use the box plot for 4 and 5.

4. What is the range of the data? _____

5. What is the interquartile range of the data?

_____

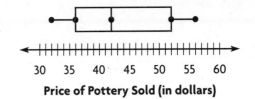

30   35   40   45   50   55   60

Price of Pottery Sold (in dollars)

Practice: Copy and Solve Find the mean absolute deviation for the data set.

6. times in minutes spent on a history quiz

35, 35, 32, 34, 34, 32, 34, 36

mean absolute deviation: _____

7. number of excused absences for one semester:

1, 2, 1, 10, 9, 9, 10, 6, 1, 1

mean absolute deviation: _____

8. The chart shows the price of different varieties of dog food at a pet store. Find the range, interquartile range, and the mean absolute deviation of the data set.

Cost of Bag of Dog Food ($)									
18	24	20	26	24	20	32	20	16	20

_____

_____

© Houghton Mifflin Harcourt Publishing Company

Problem Solving • Applications

9. **GO DEEPER** Hyato's family began a walking program. They walked 30, 45, 25, 35, 40, 30, and 40 minutes each day during one week. At the right, make a box plot of the data. Then find the interquartile range.

_____

10. **MATHEMATICAL PRACTICE 6** **Compare** Jack recorded the number of minutes his family walked each day for a month. The range of the data is 15. How does this compare to the data for Hyato's family?

_____

_____

_____

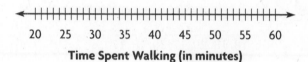

Time Spent Walking (in minutes)

11. **THINK SMARTER** **Sense or Nonsense?** Nathan claims that the interquartile range of a data set can never be greater than its range. Is Nathan's claim sense or nonsense? Explain.

_____

_____

_____

12. **THINK SMARTER** The box plot shows the heights of corn stalks from two different farms.

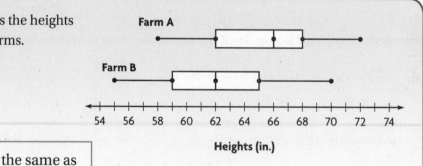

Heights (in.)

The range of Farm A's heights is [the same as / less than / greater than] the range of Farm B's heights.

© Houghton Mifflin Harcourt Publishing Company • Image Credits: (t) ©Houghton Mifflin Harcourt

Measures of Variability

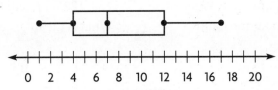

COMMON CORE STANDARD—6.SP.B.5c
Summarize and describe distributions.

1. Find the range and interquartile range of the data in the box plot.

Miles Walked

For the range, find the difference between the greatest and least values.

$\underline{\ 17\ } - \underline{\ 1\ } = \underline{\ 16\ }$

range: _____ 16 miles _____

For the interquartile range, find the difference between the upper and lower quartiles.

$\underline{\ 12\ } - \underline{\ 4\ } = \underline{\ 8\ }$

interquartile range: _____ 8 miles _____

Use the box plot for 2 and 3.

2. What is the range of the data?

_____

3. What is the interquartile range of the data?

_____

Quiz Scores

Find the mean absolute deviation for the set.

4. heights in centimeters of several flowers:

14, 7, 6, 5, 13

mean absolute deviation: _____

5. ages of several children:

5, 7, 4, 6, 3, 5, 3, 7

mean absolute deviation: _____

Problem Solving ·Real World·

6. The following data set gives the amount of time, in minutes, it took five people to cook a recipe. What is the mean absolute deviation for the data?

33, 38, 31, 36, 37

_____

7. The prices of six food processors are $63, $59, $72, $68, $61, and $67. What are the range, interquartile range, and mean absolute deviation for the data?

_____

8. **WRITE** ·*Math* Find the range, interquartile range, and mean absolute deviation for this data set: 41, 45, 60, 61, 61, 72, 80.

_____

© Houghton Mifflin Harcourt Publishing Company

Lesson Check (6.SP.B.5c)

1. Daily high temperatures recorded in a certain city are 65°F, 66°F, 70°F, 58°F, and 61°F. What is the mean absolute deviation for the data?

2. Eight different cereals have 120, 160, 135, 144, 153, 122, 118, and 134 calories per serving. What is the interquartile range for the data?

Spiral Review (6.SP.B.4, 6.SP.B.5c)

3. Look at the histogram. How many days did the restaurant sell more than 59 pizzas?

4. Look at the histogram. Where does a peak in the data occur?

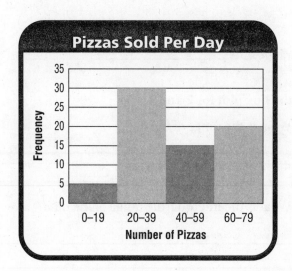

5. What is the mode of the data set?

   14, 14, 18, 20

6. The data set below lists the ages of people on a soccer team. The mean of the data is 23. What is the mean absolute deviation?

   24, 22, 19, 19, 23, 23, 26, 27, 24

© Houghton Mifflin Harcourt Publishing Company

FOR MORE PRACTICE
GO TO THE
Personal Math Trainer

Mid-Chapter Checkpoint

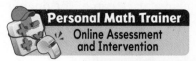
Personal Math Trainer
Online Assessment
and Intervention

Vocabulary

Vocabulary
box plot
interquartile range
mean absolute
deviation
measure of variability
range

Choose the best term from the box to complete the sentence.

1. The _____ is the difference between the upper quartile and the lower quartile of a data set. (p. 726)

2. A graph that shows the median, quartiles, and least and greatest values of a data set is called a(n) _____. (p. 714)

3. The difference between the greatest value and the least value in a data set is the _____. (p. 726)

4. The _____ is the mean of the distances between the values of a data set and the mean of the data set. (p. 720)

Concepts and Skills

5. Make a box plot for this data set: 73, 65, 68, 72, 70, 74. (6.SP.B.4)

Find the mean absolute deviation of the data. (6.SP.B.5c)

6. 43, 46, 48, 40, 38

7. 26, 20, 25, 21, 24, 27, 26, 23

8. 99, 70, 78, 85, 76, 81

_____

Find the range and interquartile range of the data. (6.SP.B.5c)

9. 2, 4, 8, 3, 2

10. 84, 82, 86, 87, 88, 83, 84

11. 39, 22, 33, 45, 42, 40, 28

_____

© Houghton Mifflin Harcourt Publishing Company

12. Yasmine keeps track of the number of hockey goals scored by her school's team at each game. The dot plot shows her data.

Goals Scored

Where is there a gap in the data? (6.SP.B.5c)

_____

13. What is the interquartile range of the data shown in the dot plot with Question 12? (6.SP.B.5c)

_____

14. `GO DEEPER` Randall's teacher added up the class scores for the quarter and used a histogram to display the data. How many peaks does the histogram have? Explain how you know. (6.SP.B.5c)

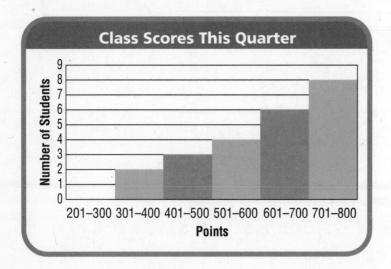

_____

15. In a box plot of the data below, where would the box be drawn? (6.SP.B.4)

55, 37, 41, 62, 50, 49, 64

_____

© Houghton Mifflin Harcourt Publishing Company

Name _____

Choose Appropriate Measures of Center and Variability

Common Core — Statistics and Probability—
6.SP.B.5d
MATHEMATICAL PRACTICES
MP2, MP4, MP6

Essential Question How can you choose appropriate measures of center and variability to describe a data set?

Outliers, gaps, and clusters in a set of data can affect both the measures of center and variability. Some measures of center and variability may describe a particular set of data better than others.

Unlock the Problem Real World

Thomas is writing an article for the school newsletter about a paper airplane competition. In the distance category, Kara's airplanes flew 17 ft, 16 ft, 18 ft, 15 ft, and 2 ft. Should Thomas use the mean, median, or mode to best describe Kara's results? Explain your reasoning.

Find the mean, median, and mode and compare them.

Mean = $\dfrac{\boxed{} + \boxed{} + \boxed{} + \boxed{} + \boxed{}}{\boxed{}}$

$= \dfrac{\boxed{}}{\boxed{}} = \boxed{}$

Order the data from least to greatest to find the median.

_____, _____, _____, _____, _____

Median = _____

The data set has no repeated values so there is no _____.

The mean is _____ than 4 of the 5 values, so it is not a good

description of the center of the data. The _____ is closer to most of the values, so it is the best way to describe Kara's results.

So, Thomas should use the _____ to describe Kara's results.

• Do you need to order the numbers?

_____

_____

_____

_____

Math Idea

The measures of center for some data sets may be very close together. If that is the case, you can list more than one measure as the best way to describe the data.

1. Explain why the two modes may be a better description than the mean or median of the data set 2, 2, 2, 2, 7, 7, 7, 7.

_____

_____

© Houghton Mifflin Harcourt Publishing Company

Example Mr. Tobin is buying a book online. He compares
prices of the book at several different sites. The table shows his results.
Make a box plot of the data. Then use the plot to find the range and
interquartile range. Which measure better describes the data? Explain
your reasoning.

Prices of Book	
Site	Price ($)
1	15
2	35
3	17
4	18
5	5
6	16
7	17

STEP 1 Make a box plot.

Write the data in order from least to
greatest.

_____, _____, _____, _____,

_____, _____, _____

Find the median of the data.

median = _____

Find the lower quartile—the median
of the lower half of the data.

lower quartile = _____

Find the upper quartile—the median
of the upper half of the data.

upper quartile = _____

Make the plot.

3 5 7 9 11 13 15 17 19 21 23 25 27 29 31 33 35 37 39
Prices of Books (in dollars)

> **Math Talk**
>
> MATHEMATICAL PRACTICES ②
>
> **Reasoning** Describe a data
> set for which the range is a
> better description than the
> interquartile range.

STEP 2 Use the box plot to find the range and the interquartile range.

range = _____ – _____ = _____

interquartile range = _____ – _____ = _____

_____ of the seven prices are within the _____.
The other two prices are much higher or lower.

So, the _____ better describes the data because the

_____ makes it appear that the data values vary more than
they actually do.

2. [THINK SMARTER] How can you tell from the box plot how varied the
data are? Explain.

_____

_____

© Houghton Mifflin Harcourt Publishing Company

Name _____

1. The distances in miles students travel to get to school are 7, 1, 5, 9, 9, and 8. Decide which measure(s) of center best describes the data set. Explain your reasoning.

mean = _____

median = _____

mode = _____

The _____ is less than 4 of the 6 data points, and the _____ describes only 2 of

the greatest data points. So, the _____ best describes the data.

2. (MATHEMATICAL PRACTICE ④) **Use Graphs** The numbers of different brands of orange juice carried in several stores are 2, 1, 3, 1, 12, 1, 2, 2, and 5. Make a box plot of the data and find the range and interquartile range. Decide which measure better describes the data set and explain your reasoning.

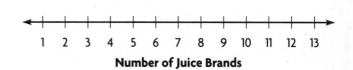

Number of Juice Brands

range = _____

interquartile range = _____

_____

_____

Math Talk — MATHEMATICAL PRACTICES ⑥
Explain How does an outlier affect the range of a data set?

On Your Own

3. (MATHEMATICAL PRACTICE ②) **Use Reasoning** The ages of students in a computer class are 14, 13, 14, 15, 14, 35, 14. Decide which measure of center(s) best describes the data set. Explain your reasoning.

mean = _____

median = _____

mode = _____

_____

_____

4. (GO DEEPER) Mateo scored 98, 85, 84, 80, 81, and 82 on six math tests. When a seventh math test score is added, the measure of center that best describes his scores is the median. What could the seventh test score be? Explain your reasoning.

_____

_____

© Houghton Mifflin Harcourt Publishing Company

Unlock the Problem Real World

5. THINK SMARTER Jaime is on the community swim team. The table shows the team's results in the last 8 swim meets. Jaime believes they can place in the top 3 at the next swim meet. Which measure of center should Jaime use to persuade her team that she is correct? Explain.

Swim Team Results	
Meet	**Place**
Meet 1	1
Meet 2	2
Meet 3	3
Meet 4	18
Meet 5	1
Meet 6	2
Meet 7	3
Meet 8	2

a. What do you need to find?

_____

_____

_____

b. What information do you need to solve the problem?

_____

_____

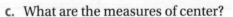

c. What are the measures of center?

_____

d. Which measure of center should Jaime use? Explain.

_____

_____

_____

Personal Math Trainer

6. THINK SMARTER ✚ The numbers of sit-ups students completed in one minute are 10, 42, 46, 50, 43, and 49. The mean of the data values is 40 and the median is 44.5. Which measure of center better describes the data, the mean or median? Use words and numbers to support your answer.

_____

_____

_____

© Houghton Mifflin Harcourt Publishing Company • Image Credits: (r) ©Digital Vision/Getty Images

Name _____

Choose Appropriate Measures of Center and Variability

Common Core **COMMON CORE STANDARD—6.SP.B.5d**
Summarize and describe distributions.

1. The distances, in miles, that 6 people travel to get to work are 14, 12, 2, 16, 16, and 18. Decide which measure(s) of center best describes the data set. Explain your reasoning.

   mean = ____13 miles____

   median = ____15 miles____

   mode = ____16 miles____

   The _____ is less than 4 of the data points, and the _____ describes only 2 of the data points. So, the _____ best describes the data.

2. The numbers of pets that several children have are 2, 1, 2, 3, 4, 3, 10, 0, 1, and 0. Make a box plot of the data and find the range and interquartile range. Decide which measure better describes the data set and explain your reasoning.

   ```
   <---+--+--+--+--+--+--+--+--+--+--+--+--+--->
       0  1  2  3  4  5  6  7  8  9  10 11 12
   ```

   range = _____ interquartile range = _____

   _____

   _____

   _____

Problem Solving Real World

3. Brett's history quiz scores are 84, 78, 92, 90, 85, 91, and 0. Decide which measure(s) of center best describes the data set. Explain your reasoning.

   mean = _____ median = _____

   mode = _____

   _____

   _____

   _____

4. Eight students were absent the following number of days in a year: 4, 8, 0, 1, 7, 2, 6, and 3. Decide if the range or interquartile range better describes the data set, and explain your reasoning.

   range = _____ interquartile range = _____

   _____

   _____

   _____

5. **WRITE** ▶ *Math* Create two sets of data that would be best described by two different measures of center.

   _____

   _____

© Houghton Mifflin Harcourt Publishing Company

Lesson Check (6.SP.B.5d)

1. Chloe used two box plots to display some data. The box in the plot for the first data set is wider than the box for the second data set. What does this say about the data?

_____

_____

_____

2. Hector recorded the temperature at noon for 7 days in a row. The temperatures are 20°F, 20°F, 20°F, 23°F, 23°F, 23°F, and 55°F. Which measure of center would best describe the data?

_____

_____

_____

Spiral Review (6.SP.B.4, 6.SP.B.5c, 6.SP.B.5d)

3. By how much does the median of the following data set change if the outlier is removed?

13, 20, 15, 19, 22, 26, 42

4. What percent of the people surveyed spent at least an hour watching television?

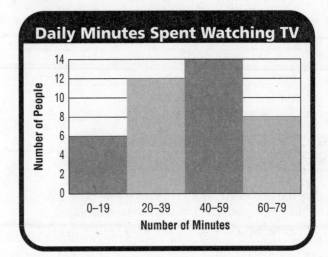

_____

5. What is the lower quartile of the following data?

12, 9, 10, 8, 7, 12

6. What is the interquartile range of the data shown in the box plot?

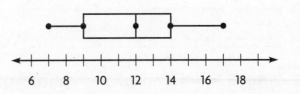

_____

© Houghton Mifflin Harcourt Publishing Company

FOR MORE PRACTICE GO TO THE Personal Math Trainer

Apply Measures of Center and Variability

Common Core Statistics and Probability— 6.SP.A.3
Also 6.SP.A.2
MATHEMATICAL PRACTICES
MP2, MP4, MP6

Essential Question What do measures of center and variability indicate about a data set?

 Unlock the Problem Real World

Julia is collecting data on her favorite sports teams for a report. The table shows the median and interquartile range of the heights of the players on her favorite baseball and basketball teams. How do the heights of the two teams compare?

Sports Team Data		
	Median	**Interquartile Range**
Baseball Team Heights	70 in.	6 in.
Basketball Team Heights	78 in.	4 in.

 Compare the medians and interquartile ranges of the two teams.

Median

The median of the _____ players' heights is _____ inches

greater than the median of the _____ players' heights.

Interquartile Range

The interquartile range of the baseball team is _____ the interquartile range of the basketball team, so the heights

of the baseball players vary _____ the heights of the basketball team.

So, the players on the _____ team are typically taller than the

players on the _____ team, and the heights of the _____

team vary more than the those of the _____ team.

Math Talk

MATHEMATICAL PRACTICES ②

Reasoning What if the mean of the heights of players on the baseball team is 75 in.? Explain what this could tell you about the data.

1. Julia randomly picks one player from the basketball team and one player from the baseball team. Given data in the table, can you say that the basketball player will definitely be taller than the baseball player? Explain your reasoning.

_____

_____

_____

© Houghton Mifflin Harcourt Publishing Company

🔒 Example Compare the means and ranges of the two data sets.

Kamira and Joey sold T-shirts during lunch to raise money for a charity. The table shows the number of T-shirts each student sold each day for two weeks. Find the mean and range of each data set, and use these measures to compare the data.

T-Shirts Sold	
Kamira	5, 1, 2, 1, 3, 3, 1, 4, 5, 5
Joey	0, 1, 2, 13, 2, 1, 3, 4, 4, 0

STEP 1 Find the mean of each data set.

Kamira:

Mean = $\dfrac{\square + \square + \square + \square + \square + \square + \square + \square + \square + \square}{\square}$

$= \dfrac{\square}{\square} = \square$

Joey:

Mean = $\dfrac{\square + \square + \square + \square + \square + \square + \square + \square + \square + \square}{\square}$

$= \dfrac{\square}{\square} = \square$

> ⚠ **ERROR Alert**
>
> Make sure you include zeroes when you count the total number of data values.

STEP 2 Find the range of each data set.

Kamira:

Range = $\square - \square = \square$

Joey:

Range = $\square - \square = \square$

STEP 3 Compare the mean and range.

The mean of Joey's sales is _____ the mean of Kamira's sales.

The range of Joey's sales is _____ the range of Kamira's sales.

So, the typical number of shirts Joey sold each day was _____ the typical number of shirts Kamira sold. However, since the range of Joey's

data was _____ than Kamira's, the number of shirts Joey sold

varied _____ from day to day than the number of shirts Kamira sold.

---

2. **MATHEMATICAL PRACTICE 6** **Explain** Which measure of center would better describe Joey's data set? Explain.

_____

_____

© Houghton Mifflin Harcourt Publishing Company • Image Credits: (tr) ©Ocean/Corbis

Name _____

1. Zoe collected data on the number of points her favorite basketball players scored in several games. Use the information in the table to compare the data.

   The mean of Player 1's points is _____ the mean of Player 2's points.

   The interquartile range of Player 1's points is _____ the interquartile range of Player 2's points.

   So, Player 2 typically scores _____ points than Player 1, but

   Player 2's scores typically vary _____ Player 1's scores.

Points Scored		
	Mean	Interquartile Range
Player 1	24	8
Player 2	33	16

2. Mark collected data on the weights of puppies at two animal shelters. Find the median and range of each data set, and use these measures to compare the data.

   _____

   _____

   _____

   _____

Puppy Weight, in pounds
Shelter A: 7, 10, 5, 12, 15, 7, 7
Shelter B: 4, 11, 5, 11, 15, 5, 13

On Your Own

Kwan analyzed data about the number of hours musicians in her band practice each week. The table shows her results. Use the table for Exercises 3–5.

3. Which two students typically practiced the same amount each week, with about the same variation in practice times?

   _____

4. Which two students typically practiced the same number of hours, but had very different variations in their practice times?

   _____

5. Which two students had the same variation in practice times, but typically practiced a different number of hours per week?

   _____

Hours of Practice per Week		
	Mean	Range
Sally	5	2
Matthew	9	12
Tim	5	12
Jennifer	5	3

© Houghton Mifflin Harcourt Publishing Company • Image Credits: (br) ©Stockbyte/Getty Images

Problem Solving • Applications (Real World)

6. **MATHEMATICAL PRACTICE ⑥ Compare** The table shows the number of miles Johnny ran each day for two weeks. Find the median and the interquartile range of each data set, and use these measures to compare the data sets.

Miles Run
Week 1   2, 1, 5, 2, 3, 3, 4
Week 2   3, 8, 1, 8, 1, 3, 1

_____

_____

_____

7. **THINK SMARTER** **Sense or Nonsense?** Yashi made the box plots at right to show the data he collected on plant growth. He thinks that the variation in bean plant growth was about the same as the variation in tomato plant growth. Does Yashi's conclusion make sense? Why or why not?

_____

_____

_____

_____

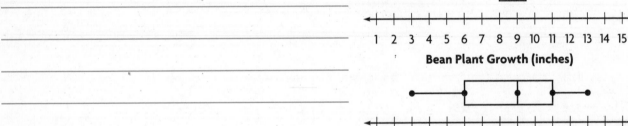

Bean Plant Growth (inches)

Tomato Plant Growth (inches)

Personal Math Trainer

8. **THINK SMARTER +** Kylie's teacher collected data on the heights of boys and girls in a sixth-grade class. Use the information in the table to compare the data.

Heights (in.)							
Girls	55	60	56	51	60	63	65
Boys	72	68	70	56	58	62	64

The mean of the boys' heights is | the same as / less than / greater than | the mean of the girls' heights.

The range of the boys' heights is | the same as / less than / greater than | the range of the girls' heights.

© Houghton Mifflin Harcourt Publishing Company

Name _____

Apply Measures of Center and Variability

COMMON CORE STANDARD—6.SP.A.3
Develop understanding of statistical variability.

Solve.

1. The table shows temperature data for two cities. Use the information in the table to compare the data.

   The mean of City 1's temperatures is ____less than____ the mean of City 2's temperatures.

   The ____interquartile range____ of City 1's temperatures is ____less than____ the ____interquartile range____ of City 2's temperatures.

   So, City 2 is typically ____warmer than____ City 1, but City 2's temperatures vary ____more than____ City 1's temperatures.

Daily High Temperatures (°F)		
	Mean	Interquartile Range
City 1	60	7
City 2	70	15

2. The table shows weights of fish that were caught in two different lakes. Find the median and range of each data set, and use these measures to compare the data.

Fish Weight (pounds)
Lake A: 7, 9, 10, 4, 6, 12
Lake B: 6, 7, 4, 5, 6, 4

_____

_____

_____

_____

Problem Solving `Real World`

3. Mrs. Mack measured the heights of her students in two classes. Class 1 has a median height of 130 cm and an interquartile range of 5 cm. Class 2 has a median height of 134 cm and an interquartile range of 8 cm. Write a statement that compares the data.

   _____

   _____

   _____

4. Richard's science test scores are 76, 80, 78, 84, and 80. His math test scores are 100, 80, 73, 94, and 71. Compare the medians and interquartile ranges.

   _____

   _____

   _____

5. **WRITE** ▸*Math* Write a short paragraph to a new student that explains how you can compare data sets by examining the mean and the interquartile range.

   _____

   _____

© Houghton Mifflin Harcourt Publishing Company

Lesson Check (6.SP.A.3)

1. Team A has a mean of 35 points and a range of 8 points. Team B has a mean of 30 points and a range of 7 points. Write a statement that compares the data.

   _____

   _____

   _____

2. Jean's test scores have a mean of 83 and an interquartile range of 4. Ben's test scores have a mean of 87 and an interquartile range of 9. Compare the students' scores.

   _____

   _____

   _____

Spiral Review (6.SP.A.3, 6.SP.B.4, 6.SP.B.5d)

3. Look at the box plots below. What is the difference between the medians for the two groups of data?

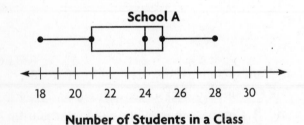

School A

Number of Students in a Class

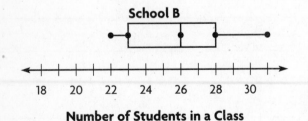

School B

Number of Students in a Class

4. The distances in miles that 6 people drive to get to work are 10, 11, 9, 12, 9, and 27. What measure of center best describes the data set?

   _____

5. Which two teams typically practice the same number of hours, but have very different variations in their practice times?

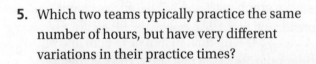

Hours of Practice Per Week		
Team	Mean	Range
A	7	1.5
B	10.5	1.5
C	7.5	5
D	10	2

_____

© Houghton Mifflin Harcourt Publishing Company

**FOR MORE PRACTICE
GO TO THE
Personal Math Trainer**

Name _____

Describe Distributions

Essential Question How can you describe the distribution of a data set collected to answer a statistical question?

Common Core Statistics and Probability—
6.SP.A.2 Also *6.SP.B.5d*
MATHEMATICAL PRACTICES
MP1, MP3, MP5, MP6

 **Activity**

Ask at least 20 students in your school how many pets they have. Record your results in a frequency table like the one shown.

Pet Survey	
Number of Pets	**Frequency**
0	
1	
2	
3	
4	

- What statistical question could you use your data to answer?

_____

_____

Unlock the Problem Real World

You can graph your data set to see the center, spread, and overall shape of the data.

Make a dot plot or a histogram of your data.

- What type of graph will you use?

_____

- How will you label your graph?

_____

_____

Math Talk

MATHEMATICAL PRACTICES ⑤

Use Tools Explain why you chose the display you used.

© Houghton Mifflin Harcourt Publishing Company

Think about the **distribution**, or overall shape of your data.

- Are there any clusters?
- Are there peaks in the data?

- Are there gaps in the data?
- Does the graph have symmetry?

1. **MATHEMATICAL PRACTICE 6** Use Math Vocabulary Describe the overall distribution of the data. Include information about clusters, gaps, peaks, and symmetry.

_____

_____

🔑 Example Find the mean, median, mode, interquartile range, and range of the data you collected.

STEP 1 Find the mean, median, and mode.

Mean: _____ Median: _____

Mode: _____

STEP 2 Draw a box plot of your data and use it to find the interquartile range and range.

Interquartile range: _____ Range: _____

---

2. Which measure of center do you think best describes your data? Why?

_____

_____

3. Does the interquartile range or range best describe your data? Why?

_____

_____

4. What is the answer to the statistical question you wrote on the previous page?

_____

_____

Math Talk

MATHEMATICAL PRACTICES 6

Describe Compare your data set to the data set of one of your classmates. How are the data sets similar and how they are different?

© Houghton Mifflin Harcourt Publishing Company

Name _____

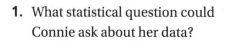 
Connie asked people their ages as they entered the food court at the mall. Use the histogram of the data she collected for 1–5.

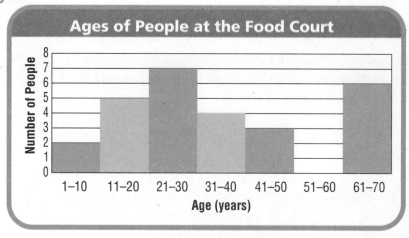

Ages of People at the Food Court

1. What statistical question could Connie ask about her data?

_____

_____

_____

2. Describe any peak or gap in the data.

_____

_____

3. Does the graph have symmetry? Explain your reasoning.

_____

_____

_____

On Your Own

4. The lower quartile of the data set is 16.5 years, and the upper quartile is 51.5 years. Find the interquartile range. Is it a better description of the data than the range? Explain your reasoning.

_____

_____

_____

_____

MATHEMATICAL PRACTICES ①

Analyze Explain what, if any, information you would need to answer the statistical question you wrote in Exercise 1 and what calculations you would need to do.

5. **MATHEMATICAL PRACTICE ③ Make Arguments** The mode of the data is 16 years old. Is the mode a good description of the center of the data? Explain.

_____

© Houghton Mifflin Harcourt Publishing Company • Image Credits: (cr) ©Lee Foster/Alamy Images

Problem Solving • Applications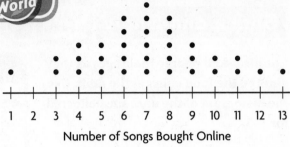

Use the dot plot for 6–8.

6. **MATHEMATICAL PRACTICE ③ Make Arguments** Jason collected data about the number of songs his classmates bought online over the past 3 weeks. Does the data set have symmetry? Why or why not?

_____

_____

Number of Songs Bought Online

7. **GO DEEPER** Jason claims that the median is a good description of his data set, but the mode is not. Does his statement make sense? Explain.

_____

_____

8. **THINK SMARTER** Trinni surveyed her classmates about how many siblings they have. A dot plot of her data increases from 0 siblings to a peak at 1 sibling, and then decreases steadily as the graph goes to 6 siblings. How is Trinni's dot plot similar to Jason's? How is it different?

_____

_____

9. **THINK SMARTER** Diego collected data on the number of movies seen last month by a random group of students.

Number of Movies Seen Last Month												
0	1	3	2	1	0	5	12	2	3	2	2	3

Draw a box plot of the data and use it to find the interquartile range and range.

Interquartile range _____

Range _____

Number of Movies Seen Last Month

© Houghton Mifflin Harcourt Publishing Company

Describe Distributions

Common Core   **COMMON CORE STANDARD—6.SP.A.2**
Develop understanding of statistical variability.

Chase asked people how many songs they have bought online in the past month. Use the histogram of the data he collected for 1–4.

1. What statistical question could Chase ask about the data?

   Possible answer: What is the median number

   of songs purchased?

2. Describe any peaks in the data.

   _____

   _____

3. Describe any gaps in the data.

   _____

   _____

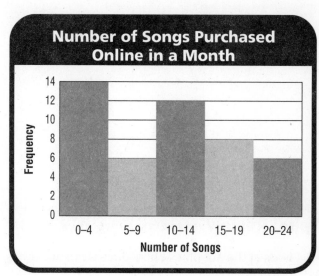

Number of Songs Purchased Online in a Month

4. Does the graph have symmetry? Explain your reasoning.

   _____

   _____

Problem Solving (Real World)

5. Mr. Carpenter teaches five classes each day. For several days in a row, he kept track of the number of students who were late to class and displayed the results in a dot plot. Describe the data.

   _____

   _____

   _____

   _____

Number of Students
Late to Class Each Day

6. **WRITE** ▸*Math* Describe how a graph of a data set can be used to understand the distribution of the data.

   _____

   _____

© Houghton Mifflin Harcourt Publishing Company

Lesson Check (6.SP.A.2)

1. The ages of people in a restaurant are 28, 10, 44, 25, 18, 8, 47, and 30. What is the median age of the people in the restaurant?

2. What is the median in the dot plot?

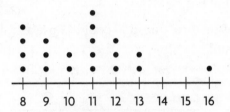

Cost (in dollars) of Dinners Ordered on Friday

Spiral Review (6.SP.A.2, 6.SP.A.3, 6.SP.B.5c, 6.SP.B.5d)

3. Look at the dot plot. Where does a gap occur in the data?

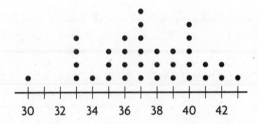

Number of Movies Ordered Per Day

4. Look at the dot plot. Where does a peak occur in the data?

5. Which two teams had similar variations in points earned, but typically earned a different number of points per game?

Points Earned Per Game		
Team	Mean	Range
Red	20	8
Blue	28	8
Green	29	4
Orange	28	4

6. Manny's monthly electric bills for the past 6 months are $140, $165, $145, $32, $125, and $135. What measure of center best represents the data?

© Houghton Mifflin Harcourt Publishing Company

FOR MORE PRACTICE
GO TO THE
Personal Math Trainer

Name _____

Problem Solving • Misleading Statistics

Essential Question How can you use the strategy *work backward* to draw conclusions about a data set?

Statistics and Probability—
6.SP.A.2 *Also 6.SP.B.5c*
MATHEMATICAL PRACTICES
MP2, MP6, MP7

 Unlock the Problem

Mr. Owen wants to move to a town where the daily high temperature is in the 70s most days. A real estate agent tells him that the mean daily high temperature in a certain town is 72°. Other statistics about the town are given in the table. Does this location match what Mr. Owen wants? Why or why not?

Use the graphic organizer to help you solve the problem.

Town Statistics for the Past Year (Daily High Temperature)	
Minimum	62°
Maximum	95°
Median	69°
Mean	72°

Read the Problem

What do I need to find?

I need to decide if the daily high temperature in the town

_____.

What information do I need?

I need the _____ in the table.

How will I use the information?

I will work backward from the statistics to draw conclusions

about the _____ of data.

Solve the Problem

The minimum high temperature is _____.

Think: The high temperature is sometimes _____ than 70°.

The maximum high temperature is _____.

Think: The high temperature is sometimes _____ than 80°.

The median of the data set is _____.

Think: The median is the middle value in the data set.

Because the median is 69°, at least half of the days must have high temperatures less than or equal to 69°.

So, the location does not match what Mr. Owen wants. The median

indicates that most days _____ have a high temperature in the 70s.

MATHEMATICAL PRACTICES ⑥

Explain Why is the mean temperature misleading in this example?

© Houghton Mifflin Harcourt Publishing Company

🔑 Try Another Problem

Ms. Garcia is buying a new car. She would like to visit a dealership that has a wide variety of cars for sale at many different price ranges. The table gives statistics about one dealership in her town. Does the dealership match Ms. Garcia's requirements? Explain your reasoning.

Statistics for New Car Prices	
Lowest Price	$12,000
Highest Price	$65,000
Lower Quartile Price	$50,000
Median Price	$55,000
Upper Quartile Price	$60,000

Read the Problem

What do I need to find?	What information do I need?	How will I use the information?

Solve the Problem

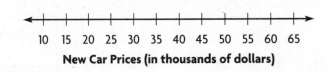

New Car Prices (in thousands of dollars)

- What would the box plot look like for a dealership that does meet Ms. Garcia's requirements?

_____

_____

_____

© Houghton Mifflin Harcourt Publishing Company

Name _____

Unlock the Problem

✓ Circle important facts.
✓ Organize the information.
✓ Choose a strategy.
✓ Check to make sure you answered the question.

1. Josh is playing a game at the carnival. If his arrow lands on a section marked 25 or higher, he gets a prize. Josh will only play if most of the players win a prize. The carnival worker says that the average (mean) score is 28. The box plot shows other statistics about the game. Should Josh play the game? Explain your reasoning.

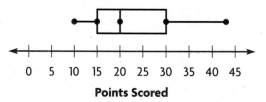

Points Scored

First, look at the median. The median is _____ points.

Next, work backward from the statistics.

The median is the _____ value of the data.

So, at least _____ of the values are scores

less than or equal to _____.

Finally, use the statistics to draw a conclusion.

_____

_____

2. THINK SMARTER **What if** a score of 15 or greater resulted in a prize? How would that affect Josh's decision? Explain.

_____

_____

3. GO DEEPER A store collects data on the sales of DVD players each week for 3 months. The manager determines that the data has a range of 62 players and decides that the weekly sales were very inconsistent. Use the statistics in the table to decide if the manager is correct. Explain your answer.

Weekly DVD Player Sales	
Minimum	16
Maximum	78
Lower quartile	58
Upper quartile	72

_____

_____

_____

© Houghton Mifflin Harcourt Publishing Company

On Your Own

4. GO DEEPER Gerard is fencing in a yard that is 21 feet by 18 feet. How many yards of fencing material does Gerard need? Explain how you found your answer.

_____

_____

5. THINK SMARTER Susanna wants to buy a fish that grows to be about 4 in. long. Mark suggests she buys the same type of fish he has. He has five of these fish with lengths of 1 in., 1 in., 6 in., 6 in., and 6 in., with a mean length of 4 in. Should Susanna buy the type of fish that Mark suggests? Explain.

_____

_____

6. MATHEMATICAL PRACTICE ⑦ **Look for a Pattern** The graph shows the number of stamps that Luciano collected over several weeks. If the pattern continues, how many stamps will Luciano collect in Week 8? Explain.

_____

_____

_____

_____

Stamps Collected

(bar graph: Number of Stamps vs Week)
Week 1: 1, Week 2: 3, Week 3: 2, Week 4: 4, Week 5: 3, Week 6: 5

7. THINK SMARTER The data set shows the number of hours Luke plays the piano each week. Luke says he usually plays the piano 3 hours per week. Why is Luke's statement misleading?

Hours Playing the Piano						
1	2	1	3	2	10	2

_____

© Houghton Mifflin Harcourt Publishing Company

Problem Solving · Misleading Statistics

Common Core COMMON CORE STANDARD—6.SP.A.2
Develop understanding of statistical variability.

Mr. Jackson wants to make dinner reservations at a restaurant that has most meals costing less than $16. The Waterside Inn advertises that they have meals that average $15. The table shows the menu items.

Menu Items	
Meal	**Price**
Potato Soup	$6
Chicken	$16
Steak	$18
Pasta	$16
Shrimp	$18
Crab Cake	$19

1. What is the minimum price and maximum price?

min = _____ **$6** _____

max = _____ **$19** _____

2. What is the mean of the prices?

_____

3. Construct a box plot for the data.

4. What is the range of the prices?

_____

5. What is the interquartile range of the prices?

_____

6. Does the menu match Mr. Jackson's requirements? Explain your reasoning.

_____

_____

_____

7. **WRITE** ▸*Math* Give an example of a misleading statistic. Explain why it is misleading.

_____

_____

© Houghton Mifflin Harcourt Publishing Company

Lesson Check (6.SP.A.2)

1. Mary's science test scores are 66, 94, 73, 81, 70, 84, and 88. What is the range of Mary's science test scores?

2. The heights in inches of students on a team are 64, 66, 60, 68, 69, 59, 60, and 70. What is the interquartile range?

Spiral Review (6.SP.B.4, 6.SP.B.5c, 6.SP.B.5d)

3. By how much does the median of the following data set change if the outlier is removed?

26, 21, 25, 18, 0, 28

4. Look at the box plot. What is the interquartile range of the data?

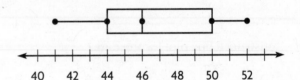

5. Erin is on the school trivia team. The table shows the team's scores in the last 8 games. Erin wants to build confidence in her team so that they will do well in the last game. If a score of 20 is considered a good score, what measure of center would be best for Erin to use to motivate her teammates?

Trivia Game Results	
Game	Score
Game 1	20
Game 2	20
Game 3	18
Game 4	19
Game 5	23
Game 6	40
Game 7	22
Game 8	19

© Houghton Mifflin Harcourt Publishing Company

FOR MORE PRACTICE
GO TO THE
Personal Math Trainer

✓ Chapter 13 Review/Test

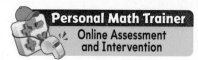

Personal Math Trainer
Online Assessment
and Intervention

1. The dot plot shows the number of chin-ups done by a gym class.

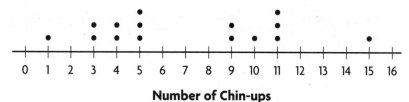

Number of Chin-ups

For numbers 1a–1e, choose Yes or No to indicate whether the statement is correct.

1a.   There are two peaks.   ○ Yes   ○ No

1b.   There are no clusters.   ○ Yes   ○ No

1c.   There is a gap from 6 to 8.   ○ Yes   ○ No

1d.   The most chin-ups anyone did   ○ Yes   ○ No
      was 15.

1e.   The modes are 3, 4, and 9.   ○ Yes   ○ No

2. The histogram shows the high temperatures in degrees Fahrenheit of various cities for one day in March.

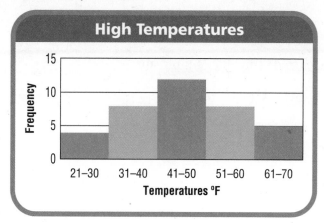

Select the best word to complete each sentence.

The histogram has | zero / one / two | peak(s). The histogram | has / does not have | symmetry.

© Houghton Mifflin Harcourt Publishing Company

GO DIGITAL   Assessment Options
Chapter Test

3. The data set shows the scores of the players on the winning team of a basketball game.

Scores of Players on Winning Team												
0	17	47	13	4	1	22	0	5	6	9	1	30

The median is
| 6. |
| 9. |
| 13. |

The lower quartile is
| 0. |
| 1. |
| 4. |

The upper quartile is
| 15 |
| 19.5 |
| 26 |

4. The data set shows the number of desks in 12 different classrooms.

Classroom Desks											
24	21	18	17	21	19	17	20	21	22	20	16

Find the values of the points on the box plot.

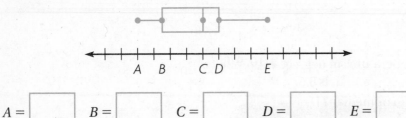

A = [ ]   B = [ ]   C = [ ]   D = [ ]   E = [ ]

5. The box plot shows the number of boxes sold at an office supply store each day for a week.

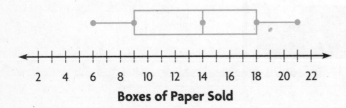

Boxes of Paper Sold

For numbers 5a–5d, select True or False for each statement.

5a. The median is 18.   ○ True   ○ False

5b. The range is 15.   ○ True   ○ False

5c. The interquartile range is 9.   ○ True   ○ False

5d. The upper quartile is 18.   ○ True   ○ False

© Houghton Mifflin Harcourt Publishing Company

Name _____

6. The data set shows the number of glasses of water Dalia drinks each day for a week.

Glasses of Water						
6	7	9	9	8	7	10

Part A

What is the mean number of glasses of water Dalia drinks each day?

_____

Part B

What is the mean absolute deviation of the number of glasses of water Dalia drinks each day? Round your answer to the nearest tenth. Use words and numbers to support your answer.

7. The numbers of emails Megan received each hour are 9, 10, 9, 8, 7, and 2. The mean of the data values is 7.5 and the median is 8.5. Which measure of center better describes the data, the mean or median? Use words and numbers to support your answer.

8. The number of miles Madelyn drove between stops was 182, 180, 181, 184, 228, and 185. Which measure of center best describes the data?

(A) mean

(B) median

(C) mode

© Houghton Mifflin Harcourt Publishing Company

9. The histogram shows the weekly earnings of part-time workers. What interval(s) represents the most common weekly earnings?

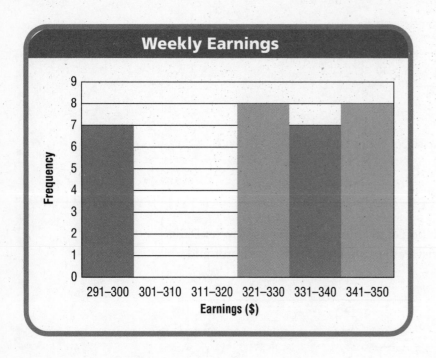

10. Jordan surveyed a group of randomly selected smartphone users and asked them how many applications they have downloaded onto their phones. The dot plot shows the results of Jordan's survey. Select the statements that describe patterns in the data. Mark all that apply.

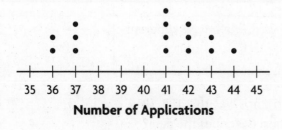

(A) The modes are 37 and 42.

(B) There is a gap from 38 to 40.

(C) There is a cluster from 41 to 44.

(D) There is a cluster from 35 to 36.

© Houghton Mifflin Harcourt Publishing Company

Name _____

11. Mrs. Gutierrez made a histogram of the birth month of the students in her class. Describe the patterns in the histogram by completing the chart.

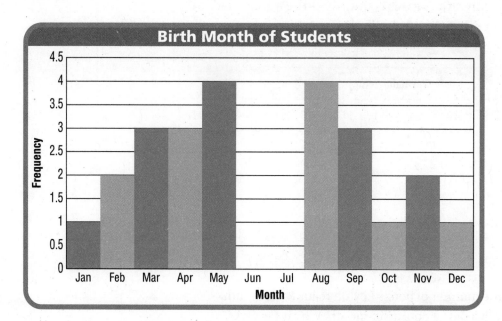

Identify any peaks.	Identify any increases across the intervals.	Identify any decreases across the intervals.

12. **GO DEEPER** Ian collected data on the number of children in 13 different families.

Number of Children												
1	2	4	3	2	1	0	8	1	1	0	2	3

Draw a box plot of the data and use it to find the interquartile range and range.

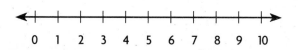

Interquartile range: _____   Range: _____

© Houghton Mifflin Harcourt Publishing Company

13.

THINK SMARTER ✚ Gavin wants to move to a county where it rains about 5 inches every month. The data set shows the monthly rainfall in inches for a county. The mean of the data is 5 and the median is 4.35. After analyzing the data, Gavin says that this county would be a good place to move. Do you agree or disagree with Gavin? Use words and numbers to support your answer.

Monthly Rainfall (in.)											
4.4	3.7	6	2.9	4.3	5.4	6.1	14.1	4.3	0.5	4.5	3.8

14. The data set shows the number of books Peyton reads each month. Peyton says she usually reads 4 books per month. Why is Peyton's statement misleading?

Books Read						
2	3	2	4	3	11	3

15. The data set shows the scores of three players for a board game.

Board Game Scores			
Player A	90	90	90
Player B	110	100	90
Player C	95	100	95

For numbers 15a–15d, choose Yes or No to indicate whether the statement is correct.

15a. The mean absolute deviation of Player B's scores is 0. ○ Yes ○ No

15b. The mean absolute deviation of Player A's scores is 0. ○ Yes ○ No

15c. The mean absolute deviation of Player B's scores is greater than the mean absolute deviation of Player C's scores. ○ Yes ○ No

© Houghton Mifflin Harcourt Publishing Company